EXTREME EATERS

BLACKBIRCH PRESS
An imprint of Thomson Gale, a part of The Thomson Corporation

Detroit • New York • San Francisco • San Diego • New Haven, Conn. • Waterville, Maine • London • Munich

© 2005 Thomson Gale, a part of The Thomson Corporation.

Thomson and Star Logo are trademarks and Gale and Blackbirch Press are registered trademarks used herein under license.

For more information, contact
Blackbirch Press
27500 Drake Rd.
Farmington Hills, MI 48331-3535
Or you can visit our Internet site at http://www.gale.com

ALL RIGHTS RESERVED.
No part of this work covered by the copyright hereon may be reproduced or used in any form or by any means—graphic, electronic, or mechanical, including photocopying, recording, taping, Web distribution or information storage retrieval systems—without the written permission of the publisher.

Every effort has been made to trace the owners of copyrighted material.

Photo credits: Cover: all pictures Corel Corporation except for top right © Royalty-Free/CORBIS; top left, bottom right © Photos.com; top center © Digital Vision; interior: all pages © Discovery Communications, Inc. except for pages 4, 12, 34 (large), 40 © Corel Corporation; page 1, 8, 16, 28, 34 (inset) © Photos.com; page 20 © Andrew Syred/Photo Researchers, Inc.; page 23 © PhotoDisc; pages 24, 32 © Royalty-Free/CORBIS; page 36 © Digital Vision

---

LIBRARY OF CONGRESS CATALOGING-IN-PUBLICATION DATA

Eaters / John Woodward, book editor.
   p. cm. — (Planet's most extreme)
  Includes bibliographical references and index.
  ISBN 1-4103-0401-9 (hardcover : alk. paper) — ISBN 1-4103-0443-4 (paper cover : alk. paper)
  1. Animals—Food—Juvenile literature. 2. Food—Juvenile literature. I. Woodward, John, 1958–. II. Title. III. Series.

  QL756.5.E18 2005
  591.5'3—dc22
                                                       2004017944

Printed in the United States of America
10 9 8 7 6 5 4 3 2 1

If you have an appetite for the unusual, you'll love what's being served up to the most extraordinary diners in the natural world. We're counting down the top ten most extreme eaters in the animal kingdom and comparing them to the unusual treats that some humans dare to eat. You could bite off more than you can chew, when eating is taken to The Most Extreme.

10

The Sloth

The sloth kicks off our countdown of extreme eaters because it eats nothing but leaves from the forests of Central and South America. It's an extremely low-energy diet, but the sloth doesn't need much energy to survive. A sloth has half as many muscles as most creatures its size. Why waste energy on muscles when you spend most of the day just hanging upside down from your fingernails?

Too lazy to hunt, the sloth likes to hang around all day in its favorite trees and munch on leaves.

A sloth doesn't waste energy looking for food, either. It only eats leaves from a very small number of tree species, which makes it a very fussy eater, unlike humans.

Right from the days of our earliest ancestors, humans have been omnivores. Dinosaurs may not have been on the menu, but we've tried eating just about everything else.

In China (top), people eat a lot more rice than Americans do. This African woman, however, is snacking on something a bit more unusual—caterpillars.

Strangely enough, humans don't instinctively know what to eat. The two things that we have in common are that most people prefer sweet things and dislike anything bitter. Everything else about our diet is picked up by copying the eating habits of the people around us. That's one reason some cultures prefer rice over bread. In China, for example, the average person eats almost eight times as much rice as the average American!

Other cultures eat things Americans see as garden pests, such as Mopani worms. In parts of Africa, these caterpillars are harvested, dried, and enjoyed as crunchy fast food.

Your parents wouldn't want you to eat like Peter Dowdeswell. He wolfs down his meals and even drinks upside down!

There's a man in England who doesn't care what he eats, just as long as he eats it quickly. Peter Dowdeswell has almost 300 world records under his belt. He's a speed eater who can devour a three-course meal in just 45 seconds! He also loves a quick drink—even if it is upside down!

Unlike Dowdeswell, the sloth will never break any records for speed eating. Thanks to its extreme diet, the sloth is undoubtedly the laziest eater in the countdown.

9 The Termite

Burrowing into number nine in the countdown is an extreme eater capable of bringing down the house. No building is safe from the insatiable appetite of the termite. It can literally eat you out of house and home. It's estimated that for every human on the planet, there is more than half a ton of termites. And they're all busily gnawing at wood.

In the wild, termites are nature's demolition crew. Their job is to search out dead, decaying trees and dispose of them. Even the tallest trees can be devoured by termites because there are just so many of them. One African termite queen can give birth to 30,000 babies every day! All those termites are fine in the wild, but we get upset when they confuse the dead wood of the forest with the dead wood in our homes!

Every year in the United States alone, these miniature home wreckers are responsible for more than 1 billion dollars worth of damage. Just one colony has been shown to eat over 150 feet of wooden boards in a year! At that rate, ten colonies would completely demolish a house in about seven years!

In nature, termites feed on dead trees. Closer to home, however, they love to munch on the wood in our houses.

Ben Thatcher has no trouble finding a snack. Whenever the toddler gets hungry, he just chews on wood.

Termites aren't the only things that can eat houses. Some kids can as well! Ben Thatcher looks like any other two year old. But he's been literally eating his mother out of house and home. Ben Thatcher is a human termite. He's been chewing on every wooden surface in the house and driving his mother crazy.

A diet of wood may be good for termites, but not for two-year-old boys. Lots of children put things in their mouths, but Ben has taken it to the extreme, which is why his worried mother sought the help of child psychologist Sandra Scott. She explains that the behavior is normal:

Children are very inquisitive and that is part of the learning process. One of the places where they have a lot of nice sensory experiences is the mouth. What does it taste like? What does it feel like? And they learn an awful lot. So children putting things in their mouth or discovering they like the sensation is completely normal.

Inside the termite's stomach we find pieces of wood and protozoa, microscopic critters that help the termite digest its woody meals.

The reason most people don't eat trees is that humans can't digest the cellulose in wood. But then, neither can termites. So how do termites manage to eat so much of something they can't actually digest? The answer is simple. They invite friends in for dinner. Termites can eat wood because their intestines are full of tiny critters called protozoa. These miniature food processors produce an enzyme that breaks down the wood into a digestible meal for the termites. The protozoa get free room and board, and the termites get to eat as much wood as they want.

8 The Macaw

The next contender in our countdown of extreme eaters can be found on a riverbank in South America. The macaw flies in at number eight because it's found an extreme way to lick a poisonous problem.

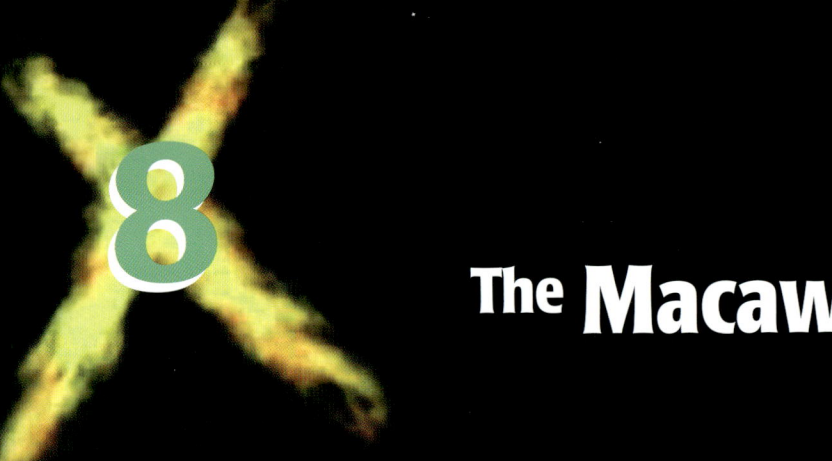

Every day, macaws eat clay. This parrot can eat a tenth of its body weight in clay each day, which is about the weight of a hamburger made of mud!

For macaws there is nothing strange about eating clay. It's both an appetizer and an antidote, for these birds are being poisoned.

Macaws love eating seeds. The trouble is that many tropical plants don't like having their seeds stolen. That's why these plants fill their seeds with incredibly poisonous alkaloids and tannins. It's a great defense, but it doesn't stop the macaw. Researchers believe that by eating clay, this parrot has found an antidote to the poisons. It's thought that the minerals in the clay bind with the seeds' toxic substances, neutralizing the poison so that it passes harmlessly through the bird's digestive system.

Macaws like to eat clay as an antidote to the poisons that are found in their favorite tropical seeds and fruits.

In Santa Monica, California, naturopathic physician Brian Roettger believes that we could learn a lot from the macaw. We don't have to visit a muddy riverbank, however. This homeopathic pharmacy stocks a most unusual remedy.

Could a teaspoon of clay keep the doctor away? The theory is that just like in the macaw, when this special bentonite clay passes through our body, it absorbs impurities like heavy metals and

A teaspoon of clay might keep the doctor away, according to those who believe clay removes impurities from our bodies.

Brian Roettger believes that drinking a bit of bentonite clay each day keeps toxins from building up in his system.

pesticides that we're exposed to every day. Some tests have shown that clay may have other beneficial effects, too. Roettger explains:

> They took some bacteria and put it in a beaker and grew it. And they put bentonite clay in that beaker. Within 90 to 100 minutes it had removed almost 100 percent of the bacteria, so it has a tremendous ability to absorb toxins.

If just one teaspoon of clay works wonders for us, just imagine the benefits for a macaw. They can eat up to twenty teaspoons of clay every day!

7 The **Parrotfish**

We know macaws eat chunks of clay, but there's also another parrot that likes its food rock hard. And if you think that sounds a little fishy, you're right. Because the next parrot is actually the parrotfish.

The parrotfish gets its name from its large front teeth, which are fused together like a parrot's beak. The parrotfish is number seven in the countdown because it uses its beak to break off tiny pieces of stony coral. The skeleton of reef-building corals is as hard as rock. A parrotfish will eat up to 4,000 mouthfuls of rock every day! It grinds up the coral using specialized teeth in the back of its throat. It's not after the rock, but the animals that made it.

The corals you see on the reef are actually houses for thousands of tiny animals. Looking like miniature sea anemones, they use their tiny tentacles to feed on plankton drifting past in the water. Coral is actually a living rock, and it's the cells of these coral polyps that are food for the parrotfish.

Just as you can't make an omelette without breaking eggs, you can't eat polyps without getting a beak full of rock. That's why every day, parrotfish can chew their way through six pounds of rock!

To reach the tasty polyps that live in the hard coral, the parrotfish will chew through pounds of the rock!

While most people could never stomach such an extreme diet, it could tempt the taste buds of one Lithuanian woman. This is the home of a human parrotfish. When Stanislava Montviliene pops out to get her lunch, she doesn't go to a supermarket. She likes to eat dirt.

Eating dirt is far more common than you'd expect. Surveys have found that as many as 50 percent of adult women in rural areas of

Stanislava Montviliene likes to eat dirt. Perhaps she isn't getting enough minerals in her diet.

The rock this parrotfish is eating will pass right through its body and come out as fine, white coral sand.

the American South had eaten at least two ounces of dirt in the previous two weeks. Some think they're satisfying a craving for minerals lacking in their normal diet. However, human parrotfish face real dangers because eating the wrong type of dirt can actually cause malnutrition.

Real parrotfish have no such problems because all the rock they eat passes straight through them and comes out as fine white sand. Each year a parrotfish can produce a ton of coral sand! So next time you're on vacation on a beach in the tropics, just remember that you're sitting in lovely white parrotfish droppings.

6 The Dust Mite

Some extreme eaters can be truly horrific. Horror movies have thrived on our nightmares about being eaten alive. But there's one animal that's never made it onto the big screen, yet it eats more of us than any other creature on Earth. Say hello to the dust mite. It's so small that you can't see it without a microscope.

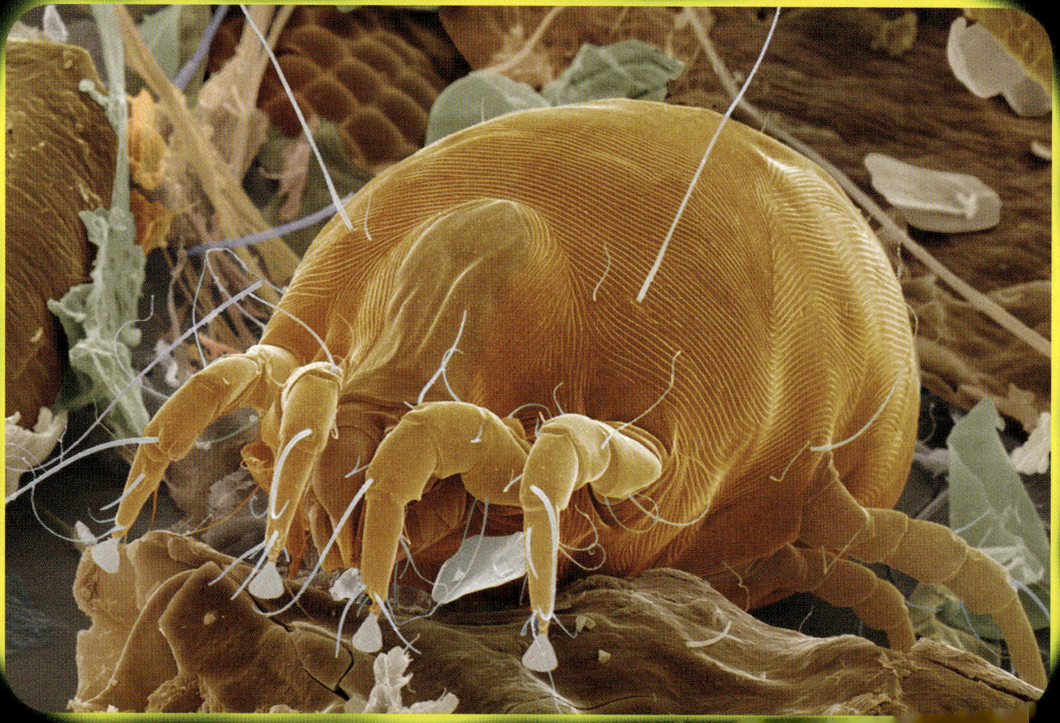

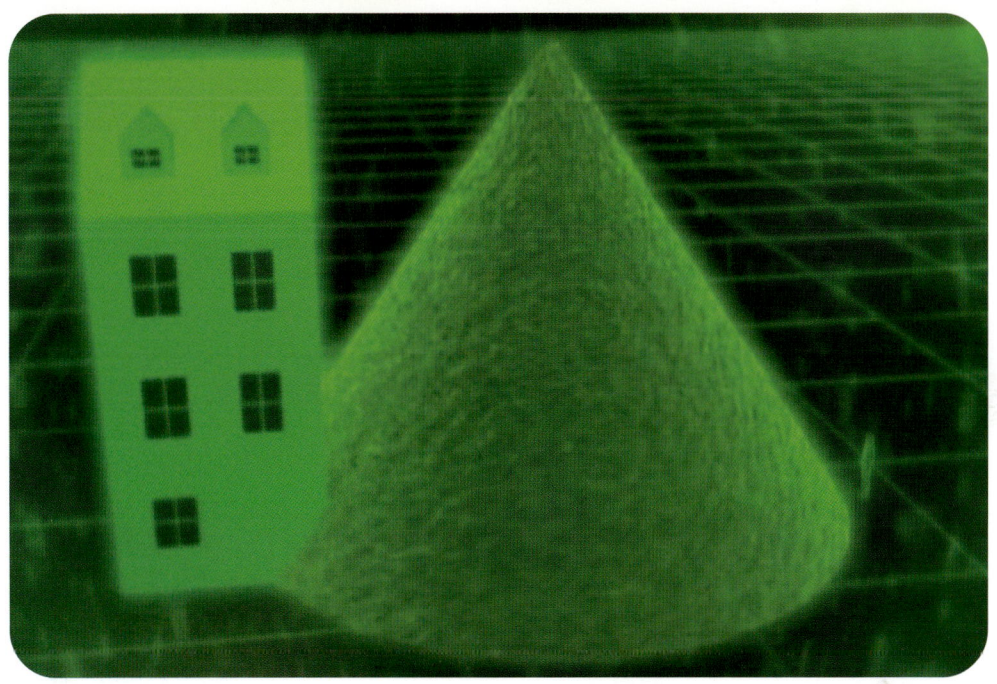

Every day, the world's population sheds enough dead skin cells to fill a four-story house.

We spend a third of our lives in bed. That means we spend a third of our lives next to the 2 million dust mites that can live in our mattresses! These eight-legged scavengers are number six in the countdown because every night they feast on our bodies.

Dust mites eat our skin. Each day, we produce an incredible number of skin cells for them to eat. In the next five seconds you will lose more than 4,000 skin cells. It's perfectly natural. It's part of your body's defense system for you to shed more than a million dead skin cells every hour. It means that in just one day, the amount of skin shed by the world's human population would fill a four-story house!

Eighty percent of the dust you see floating in a sunbeam is actually flakes of dead skin. This is food for the dust mites that live in your bed.

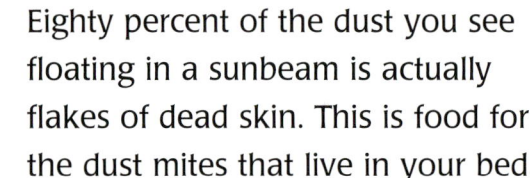

The dust particles you see in a sunbeam are mostly dead skin flakes, a feast for dust mites.

Take a close look at your bed tonight. You just might find some of the twenty or so pellets of digested skin that dust mites drop each day.

Just like in any animal, what goes in the dust mite must come out. What comes out of a dust mite is a pellet of digested skin. Each day about twenty of these tiny pellets join the piles of dust mite carcasses and cast-off skins that accumulate in your bed. So tonight when you go to bed, take another look at your pillow. If it's more than a couple of years old, about 10 percent of its weight will be from dust mites and their droppings!

5 The Cleaner Fish

The next contender in our countdown of extreme eaters knows no fear. It's a fish that'll take a bite out of anything in the water, including us! The shark may seem like an extreme eater when you're in the water with one, but there's another fish that eats shark for breakfast—the cleaner fish.

These little fish with big appetites are number five in the countdown because they eat other fish. And incredibly, the fish let them! They're possibly the bravest fish in the world, because cleaner fish boldly go where no fish have gone before—often right into the jaws of death!

A fish invites a cleaner fish to groom inside its mouth. Cleaner fish live off the parasites and dead skin of other fish.

As their name suggests, cleaner fish groom other fish by eating their decaying skin and parasites. Scientists found that a single cleaner on the Great Barrier Reef can eat 1,200 parasites in a single day, having serviced more than 2,500 fish! While it may be a lousy job, it seems there is no shortage of clients, as Paul Clarkson from California's Aquarium of the Pacific explains:

> It's a bit of a fish car wash. There'll be certain areas of the reef where cleaner fish will set up. The other animals will come into that section and it'll be kind of a no-eating zone where the larger animals won't eat these smaller fish. They'll allow them to approach and clean them.

Clients hang motionless, spreading their fins to allow cleaners access into those hard to reach places. Cleaner fish are number five in the countdown because they don't just chew on fishes. They'll clean up any skin in the neighborhood—including our own!

A fish sits still as cleaner fish go to work. How much do you tip a cleaner fish anyway?

In this Turkish spa, there's something fishy going on. People with skin conditions like eczema and psoriasis have come here to be nibbled on by three types of cleaner fish. There's a striker to eat the patient's flaky skin; a sucker to draw blood; and a healer who uses saliva to dress the wound. After two weeks of fish therapy, the results are said to be miraculous, for when it comes to getting under our skin, nothing does it better than the cleaner fish.

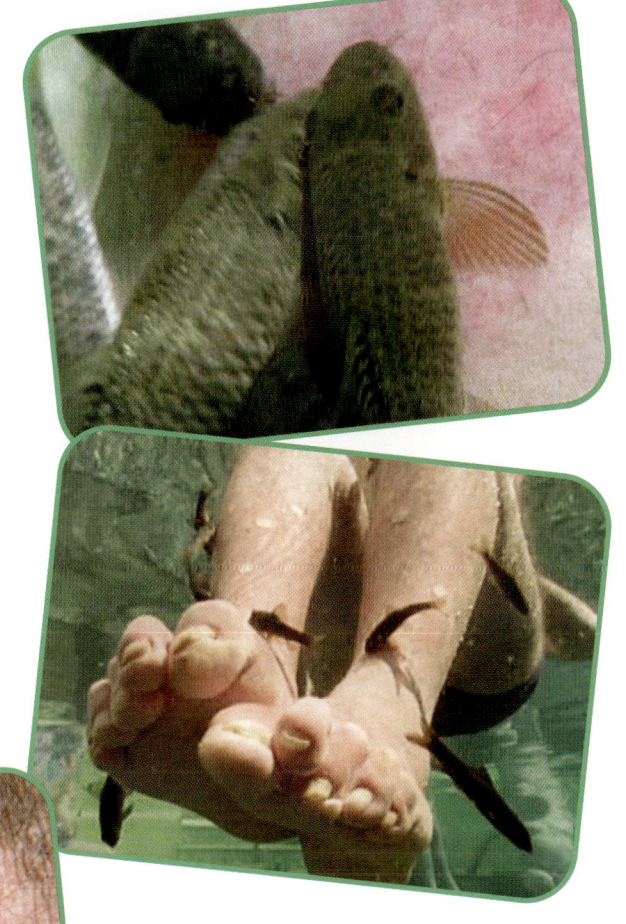

During fish therapy (from top), strikers eat the patient's flaky skin; then, suckers draw blood; and finally, a healer dresses the wound with saliva.

4

The Dung Beetle

Elephants have an extremely big appetite, which is why every day an adult elephant produces over 300 pounds of dung. Disposing of all this dung is a big job, but it's no trouble for the animal that's rolling into number four in the countdown—the dung beetle, nature's walking waste-disposal unit.

A steaming pile of fresh dung can attract up to 1,600 beetles, and these extreme eaters can remove it all in less than two hours. We may think that dung is distasteful, but for these beetles, there's fierce competition for the best bits. What can't be eaten on the spot is rolled into a ball to become a tasty takeout.

Dung beetles hide their leftovers in underground larders. It may be hard for us to imagine that a ball of dung could be this desirable, but the world is full of dung-nappers!

The beetle has another reason for burying its precious food ball. There's nothing like a ball of dung to make a safe, if somewhat smelly, nursery. In here, the baby beetle can grow up surrounded by its pre-digested food.

After feasting on elephant droppings (top), a dung beetle rolls up the leftovers and hides them (middle). Meanwhile, a baby is snug in the nursery (bottom).

This woman is taking part in a dung-spitting contest. Participants must spit pellets of antelope dung as far as they can.

A person's mouth is an unusual place to find antelope dung. But there are unusual people who actually put dung in their mouths on purpose!

Unlike the dung beetle, however, these unusual people don't swallow. They spit. This is the ancient African sport of dung spitting. For the winner, there's a prize, a kiss, and a much-needed breath mint.

Thankfully, the smell of fresh droppings appeals to the hard-working dung beetle. Could you imagine the mess that would pile up if these extreme eaters ever lost their appetite for dung?

The dung beetle deserves thanks for doing a very dirty job. Without them, we'd be up to our eyeballs in a great big mess.

3 The Poison Dart Frog

The next contender in our countdown of extreme eaters is the most poisonous animal on the planet. Hiding away at number three in the countdown is an animal with such an extreme diet that it can produce a poison 25 times more lethal than cobra venom! You'd have to be crazy to go anywhere near poison dart frogs!

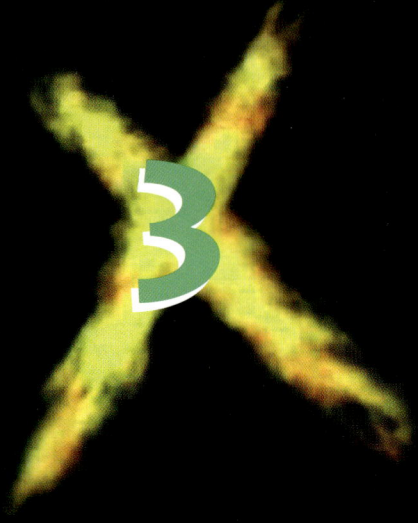

You wouldn't want to go anywhere near this poison dart frog. It stores very potent venom in its skin.

In the wild, these frogs eat ants. Somehow they convert the highly toxic alkaloids in ant venom into the poison they store in their skin. To find poisonous poison dart frogs, you need to travel to the jungles of South America. Poison dart frogs are number three in the countdown because they make use of a deadly diet. It seems that you really are what you eat.

How many bug parts are in this tasty snack? Bits of insects regularly make their way into our food as it is processed.

Frogs are not the only ones eating insects. In your lifetime you can expect to eat a pound of bugs! That's because insects like eating the same things we do. And sometimes they can get caught up in the process of food manufacturing.

Because of the way we process our food, most of us will never notice the insects that end up on our plate. But fear not! In the United States, the FDA is looking out for you. By law you'll never find more than 30 insect eggs in a can of tomato paste; 20 maggots in a can of mushrooms, or a block of chocolate that has more than 80 insect fragments. The next time you feel an annoying tickle when you swallow, the chances are high that it's not a frog in your throat.

Insects like to eat wheat (inset) as much as we do. Think about that the next time you eat bread or cereal.

2 The Meerkat

It's not easy being a meerkat. This little animal lives in one of the harshest environments on Earth—the Kalahari Desert in Africa. The meerkat lives in a land full of dangers, one of which is its deadly diet. Meerkats are number two in the countdown because they just love to eat scorpions!

Having such a deadly diet requires some special abilities. The first thing they do is try to get rid of that venomous stinger. Even if they get stung, however, it will probably survive. Meerkats have developed a certain amount of immunity to scorpion venom. This is a useful tool when you're eating one of the most venomous animals on the planet!

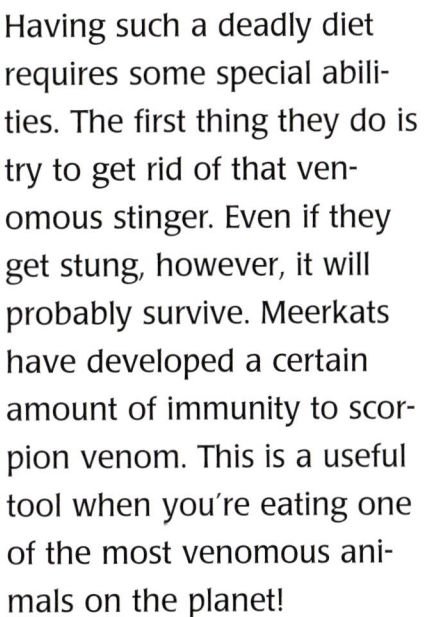

A meerkat drags a scorpion from its lair (top), before removing its poisonous stinger (close-up, bottom).

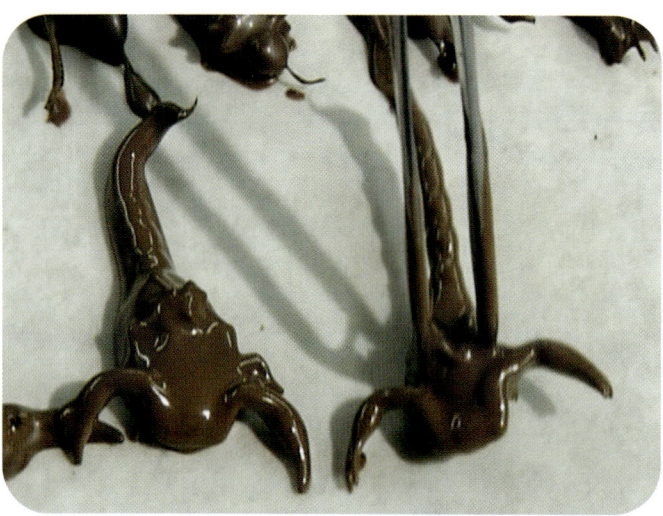

Even a chocolate addict would find it hard to swallow scorpions dipped in chocolate.

Most people would find a scorpion hard to swallow. Would it be more appetizing if it were dipped in chocolate? Desert scorpions become dessert scorpions at one of the most unusual candy stores in the world. Hotlix of Pismo Beach, California, manufactures the only candy apples on the planet where the worms are on the outside!

Ten years ago, Larry Peterman started putting worms in lollipops. What started off as a gimmick is now a flourishing business. After all, there's no reason why we can't eat bugs. He explains:

This candy maker likes to dip his candy apples in worms, and his lollipops have an insect surprise in the middle.

Insects have a lot of protein and carbohydrates— stuff that's very good for you. Not too long ago in the Bosnian war, a pilot who was downed lived on insects. People all around the world have been eating insects for years. It's just that our Western culture doesn't appreciate them as much as other cultures do.

The **Vulture**

A meerkat doesn't need a chocolate coating to sweeten its poisonous prey. And yet not even a meal of scorpion venom is as deadly as the diet of The Most Extreme eater in the countdown.

A vulture soaring in the air looks for a dead animal to eat. Vultures love stinking, rotting flesh.

The number one Most Extreme eater in the countdown has the world's deadliest diet. It eats the most lethal substances on the planet as if it were candy. Soaring into number one in the countdown is the vulture.

Thanks to its incredible nose, a vulture can literally smell death a mile away. And for a vulture, something that stinks to high heaven is a sure sign that dinner is served.

Deadly and disgusting, rotting flesh contains dangerous bacteria such as cholera and anthrax. Botulism bacteria produce the most lethal substance on the planet. Just one part per trillion is enough to kill you!

It's no wonder that humans avoid rotting flesh. At the first sign of decay we leave food alone.

A gourmet banquet for a vulture, the rotting flesh of this dead cow is full of bacteria that would kill humans.

Because their stomach acids are strong enough to kill most harmful bacteria (inset), vultures can survive their toxic meals.

The vulture, however, is number one in the countdown because it can eat rotting flesh that would kill other animals. That's because the vulture has a stomach of steel. The vulture's stomach acids are so strong that they'll destroy almost all bacteria.

Biologists have also discovered that the nervous system of the turkey vulture is resistant to botulism, and its immune system is powerful enough to deal with most bacterial toxins.

There is one thing that not even a vulture would touch, however. At Sideshows by the Seashore at Coney Island, this man is cooking up something special. Meet the Amazing, Blazing Tyler Fyre! He says fire eating is not a trick:

> Fire eating is an exciting act that everyone wants to learn. It is a lot of fun to perform. People throughout the world are convinced that it's cold fire, or trick fire, or canned fire. There are people who say it's a special gel you put on your mouth beforehand, and there are a lot of days that I wish that were true. But if there were such a

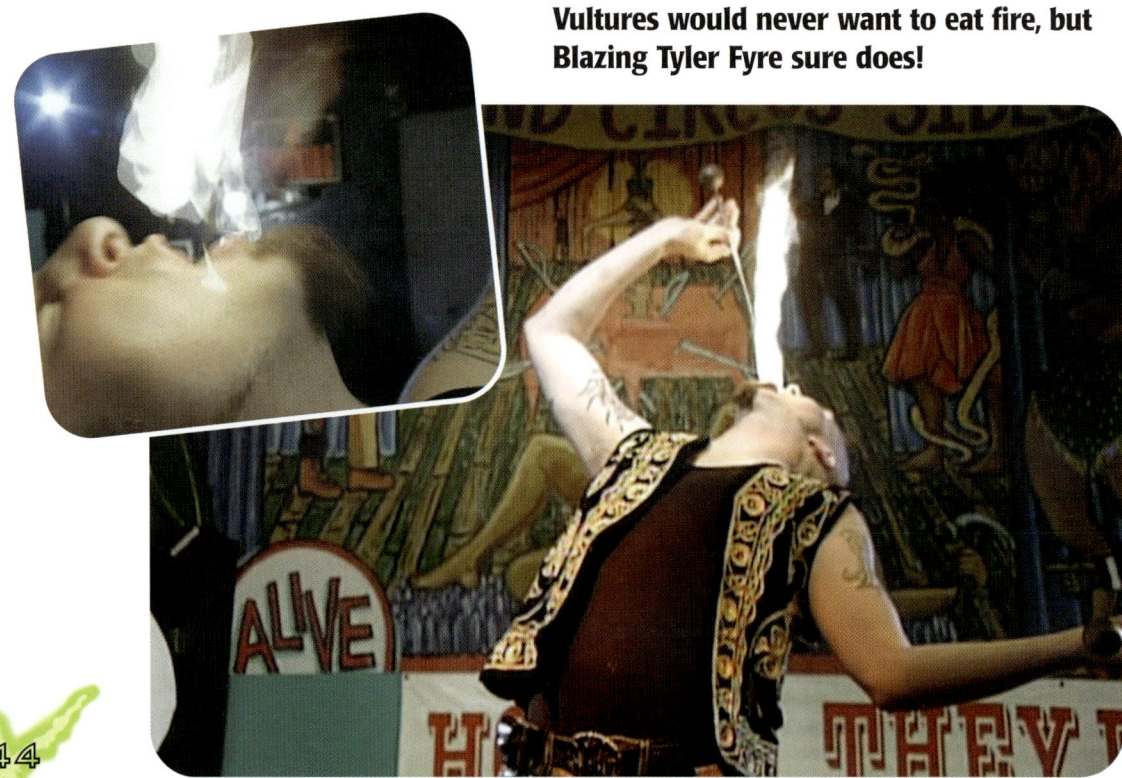

Vultures would never want to eat fire, but Blazing Tyler Fyre sure does!

Vultures enjoy a barbecued meal after a fire. Vultures are not very picky eaters—they'll eat flesh any way it's served.

> burnproof, flameproof gel, firemen wouldn't have to wear yellow suits anymore. So while fire eating is not a trick, there is a way to control and manipulate the fire that keeps it from actually burning the flesh.

Vultures may not eat fire, but they do enjoy a good barbecue. Cooked, raw, or rotting, vultures will eat any flesh.

A diet this disgusting is enough to make you sick. It can even make a vulture throw up. But that's only because it has eaten so much that it needs to lighten the load so it can get off the ground! With table manners like this, it's no wonder that when it comes to eating, the vulture really is The Most Extreme.

For More Information

Linda Jacobs Altman, *Parrots*. Tarrytown, NY: Marshall Cavendish, 2000.

George Barlow, *The Cichlid Fishes: Nature's Grand Experiment in Evolution*. New York: Perseus, 2002.

Andrew Hipp, *Dung Beetles*. New York: Rosen, 2003.

Jinny Johnson, *Vultures*. Oxford: Raintree, 2003.

Deborah Kops, *Vultures*. San Diego: Blackbirch Press, 2000.

Joy Paige, *Sloth: The World's Slowest Mammal*. New York: Rosen, 2002.

Lola M.M. Schaefer, *Parrotfish*. Mankato, MN: Capstone Press, 2000.

Mari C. Schuh, *Termites*. Mankato, MN: Capstone Press, 2003.

Christy Steele, *Sloths*. New York: Harcourt, 2002.

Glossary

alkaloids: bitter, organic materials produced by some seed plants

anthrax: a disease that occurs in animals and may be fatal to humans

antidote: a remedy to counteract poison

botulism: a disease that causes paralysis

cholera: an intestinal disease that often causes severe diarrhea

coral: the hard skeleton produced by a sea creature called a polyp

enzyme: a protein produced by living cells that helps certain biological processes

malnutrition: the state of poor health produced by poor diet

naturopathic: an alternative medical treatment that avoids drugs and surgery

omnivore: animal that eats both animal and vegetable materials

protozoa: a class of microscopic organisms that includes some parasites

tannin: a harsh substance derived from some plants

toxic: capable of causing injury or death

venom: a poison produced by some animals and insects

Index

ants, 33

babies, 9, 29
bacteria, 15, 42, 43
beak, 17
bed, 21–23
beetle, 28–31
botulism, 42, 43

candy, 38–39
caterpillars, 6
cellulose, 11
children, 10–11
clay, 13–15
cleaner fish, 24–27
coral, 17

dart frog, 32–35
death, 41
digestion, 11, 13
dirt, 18–19
droppings, 19, 23, 28, 31
dung beetle, 28–31
dust mite, 20–23

elephants, 28

FDA, 35
fire eating, 44–45
fish, 16–19, 24–27
flesh, rotting, 42–43, 45
food processing, 35
frog, 32–35

humans, 5–6

insects, 34–35, 39

leaves, 4–5

macaw, 12–15
mattress, 21
meerkat, 36–39
muscles, 4

nose, 41

omnivores, 5

parasites, 25
parrotfish, 16–19
poison, 13–15, 42–43
poison dart frog, 32–35
protozoa, 11

rice, 6

sand, 19
scorpions, 36–38
seeds, 13
skin, 21–22, 25, 26–27
sloth, 4–7
stinger, 37

termite, 8–11
trees, decaying, 9, 11

venom, 33, 37
vulture, 40–45

wood, 8–11
worms, 38